Richard Mosqueda

Digital protection relay for a gas turbine generator

Richard Mosqueda

Digital protection relay for a gas turbine generator

A detailed explanation of the calculations of protection settings selected to protect a generator.

ScienciaScripts

Imprint
Any brand names and product names mentioned in this book are subject to trademark, brand or patent protection and are trademarks or registered trademarks of their respective holders. The use of brand names, product names, common names, trade names, product descriptions etc. even without a particular marking in this work is in no way to be construed to mean that such names may be regarded as unrestricted in respect of trademark and brand protection legislation and could thus be used by anyone.

Cover image: www.ingimage.com

This book is a translation from the original published under ISBN 978-620-0-04661-1.

Publisher:
Sciencia Scripts
is a trademark of
Dodo Books Indian Ocean Ltd. and OmniScriptum S.R.L publishing group

120 High Road, East Finchley, London, N2 9ED, United Kingdom
Str. Armeneasca 28/1, office 1, Chisinau MD-2012, Republic of Moldova, Europe
Printed at: see last page
ISBN: 978-620-7-15637-5

Digital protection relay for a gas turbine generator

Carried out by: Eng°. Richard Mosqueda R.
richardmosqueda@gmail.com

Introduction:

The cement plant "Venezolana de Cementos" (formerly Cemex Venezuela) located in Pertigalete - Venezuela, has an installed capacity of 80 MW and a nominal operating power of approximately 54 MW. The electrical system of this plant is isolated to ground at 13800 V, which means that ground faults are of low magnitude (approx. 10 amps) and the voltage to ground is very high (approx. 7.9 Kv).

This cement plant has a power generation plant with 4 gas turbines: GT-8 (30 MW), GT-7 (30 MW), GT-6 (15 MW), GT-5 (7 MW) and its protection systems are from the 1970s and need to be updated.

This article explains the replacement of the existing GT-7 Gas Turbine generator protection system with a multifunction generator protection relay: SIEMENS, Type: 7UM62255EB093CG0 with a 7XT71 controller unit, 7XR6004 resistor unit for rotor faults, and a 3PP1326 resistor unit for excitation faults.

A step-by-step description of the functions and configuration of the protection equipment will be given, with a brief explanation of installation and commissioning.

TABLE OF CONTENTS

Chapter 1 — 8

Chapter 2 — 14

Chapter 3 — 32

Chapter 4 — 34

Chapter 5 — 41

Current protections of the GT-7 Generator:

ÑAME	IEC	ANSI	CONSTR	BRAND & TYPE
ABOUT VOLTAGE	$U>$	59	ELECTRO-M	BBC - UMX31
ENERGY RETURN	P	32	ELECTRO-M	BBC - PM2g90-110
GROUNDING		67 \	ELECTRO-M	BBC - PM2ÍÍ26-108
EXCITATION FAULT	$\varphi<$	40	ELECTRO-M	BBC-YP/WUX101 + BBC - PUM21-110
ROTOR GROUNDING	$Re<$	B4R	ELECTRO-M	BBC - PUM20H45-110
OVERCURRENT	$l>/l>>>$	5M1	ELECTRO -M	BBC-SW
ASYMMETRIC LOADING	$I2$	46	ELECTRO -M	BBC - IG25-2 + BBC - CSM2Cd
DIFFERENTIAL	$l\Delta$	87G	ELECTRO-M	BBC - D2Se1
LOW FREQUENCY	$f<$	81	ELECTRONIC	BBC - FCX103b/1

GT-7 Generator Current Protections Diagram:

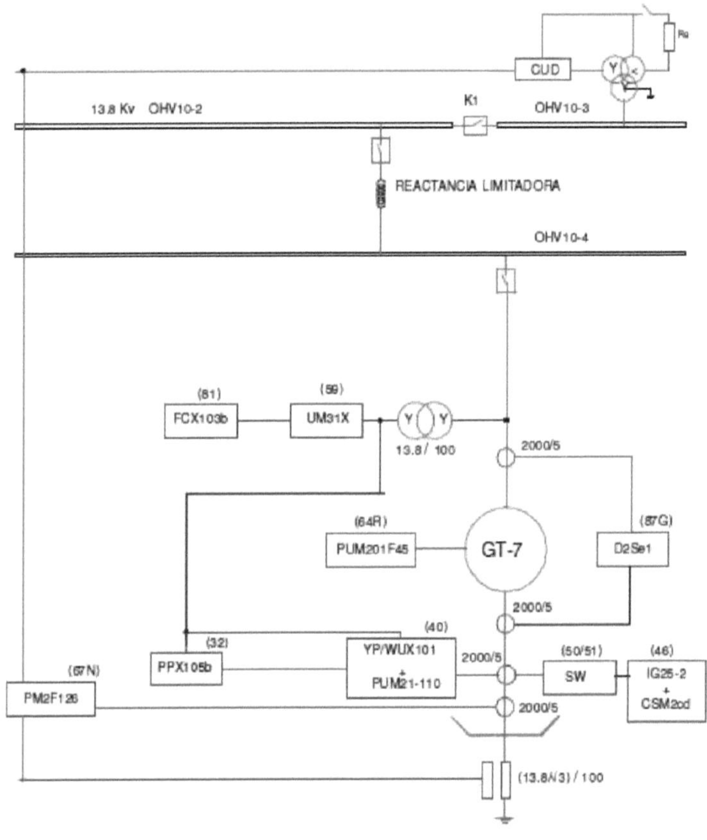

GT-7 Generator connection diagram:

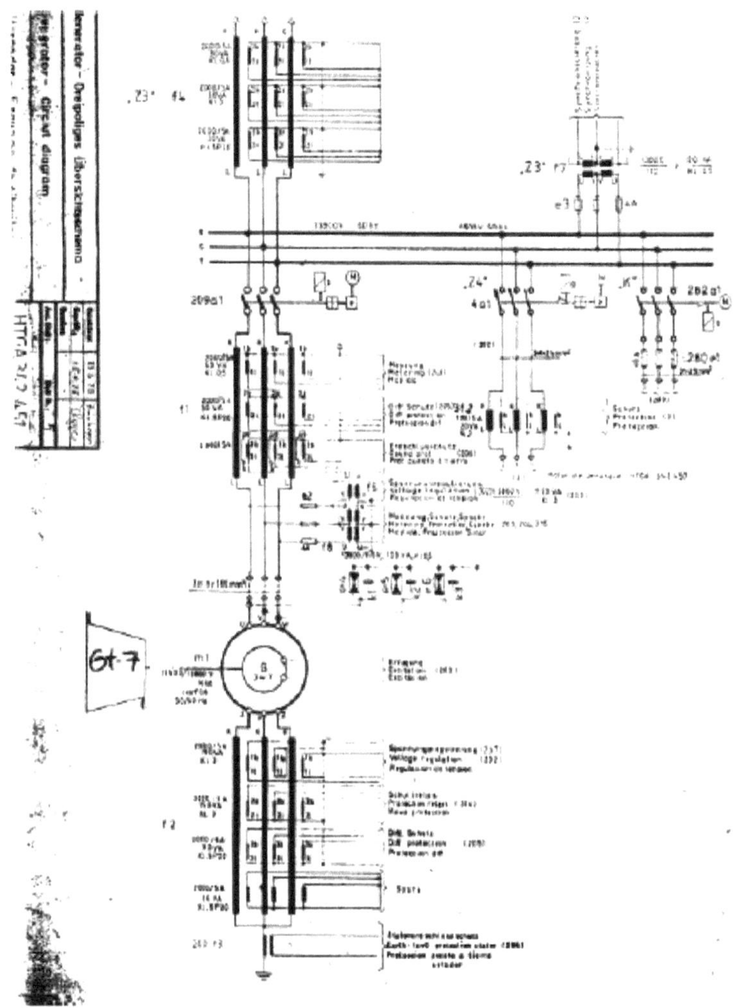

Sequence of current alarms and tripping of the GT-7 Generator protections:

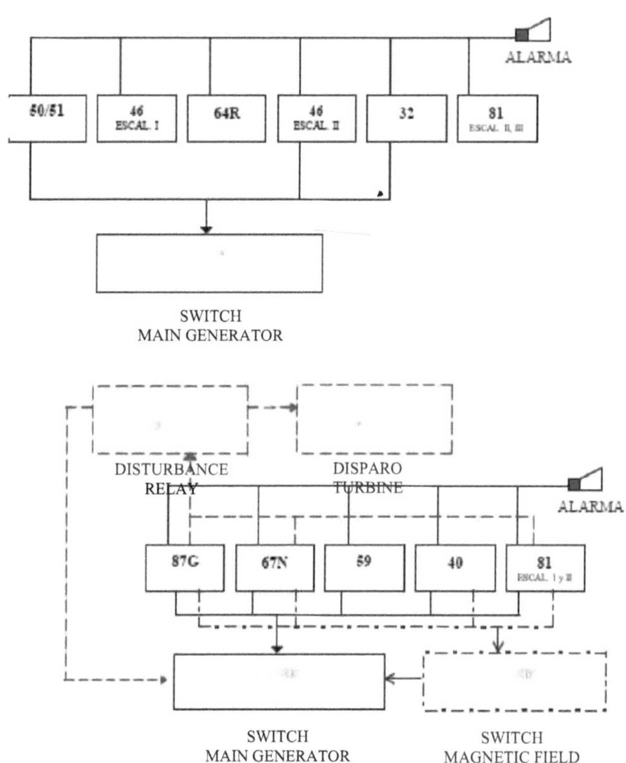

SIEMENS protection relay settings - 7UM62255EB093CG0:

Using the DIGSI 4 software, we go to **Plant Data**:

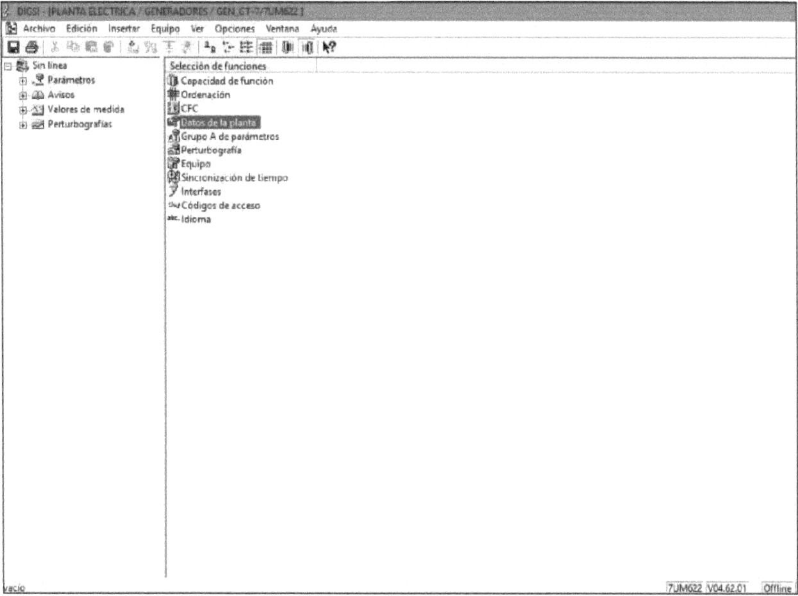

1.   - Plant data:

1.1.- Network Data:

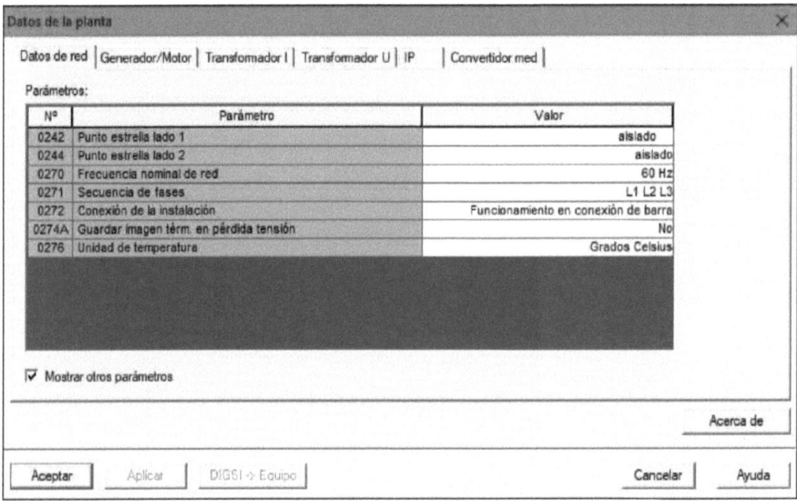

0242 Star point side 1 and **0244** Star point side 2 should be placed as: "isolated", since as mentioned before the electrical system is isolated at 13.8 Kv level.

In **0272** System configuration: "Busbar connection operation" is selected since the GT-7 Generator is connected directly to the busbar and is not connected to a transformer.

1.2.- Generator/Motor:

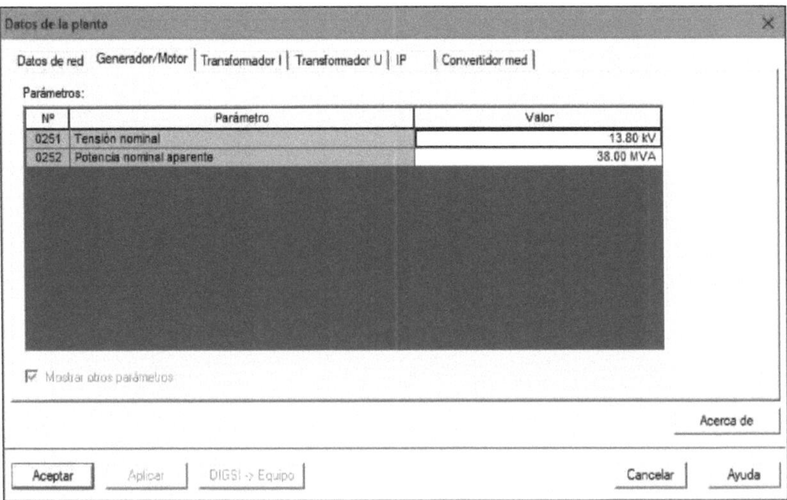

From the electrical data of the generator:

0251 Rated voltage: 13800 volts

0252 Apparent rated power: 38 MVA

1.3.- Transformer I:

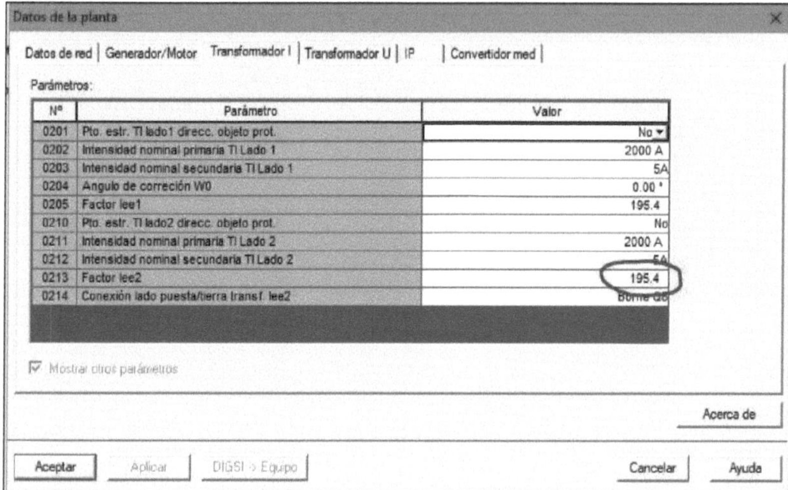

CT side 1: 2000/5, CT side 2: 2000/5, are entered at addresses **0202** and **0211.**
At address **0213** the Iee2 Factor, but the value of TC Iee2 is not available. It was necessary to inject primary current and measure at points (p-q of 200f1) with an ammeter:

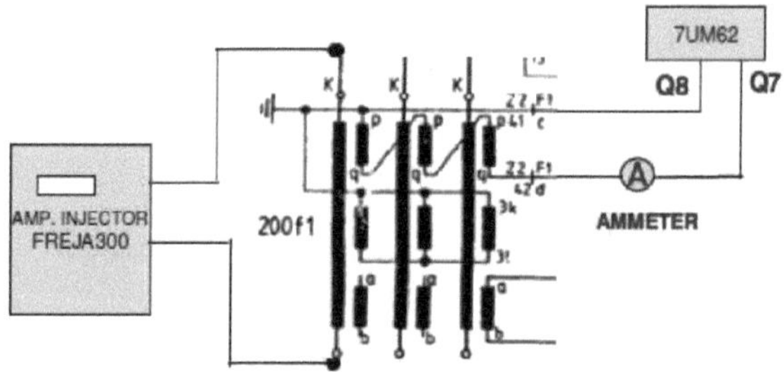

PRIMARY INJECTION VALUE (Amp)	VALUE MEASURED IN AMMETER and SIPROTEC (mA)
10	51.19
8	40.94
6	30.70
4	10.24

We calculate:

R10 =10 / 0.05119=195. 4A

    R8=8/0.04094=195.4A

    R6=6/0.0307=195. 4A

    R4=4/0.02047=195.4A

Then,

0213 Iee2 factor: 195.4

In **0205**, the value entered is 195.4

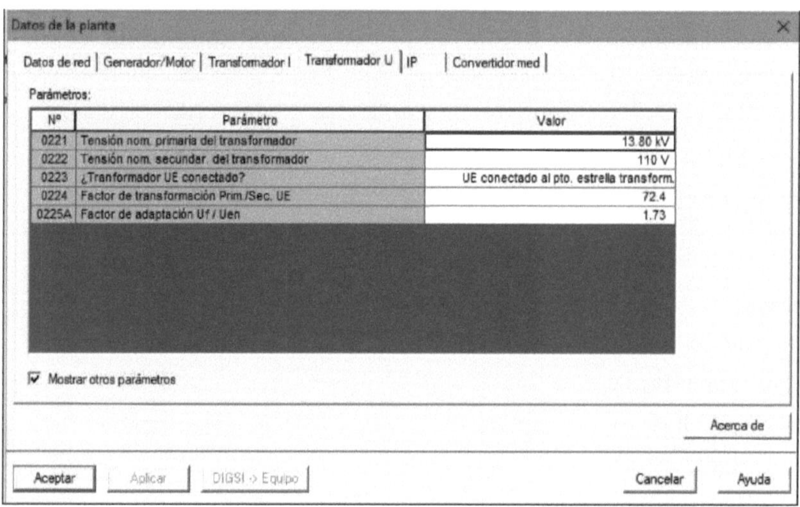

The EU Factor:

$$Factor\ UE = \frac{Utransf\ prim}{UE\ sec} = \frac{\frac{13800}{\sqrt{3}}}{110} = 72.43$$

then: **0224** Transformation factor Prim / Sec. UE = 72.4

From: (Uph / Udelta) = 3/√3 = 1.73

Then:

0225A Adaptation factor Uf / Uen = 1.73

Converter med:

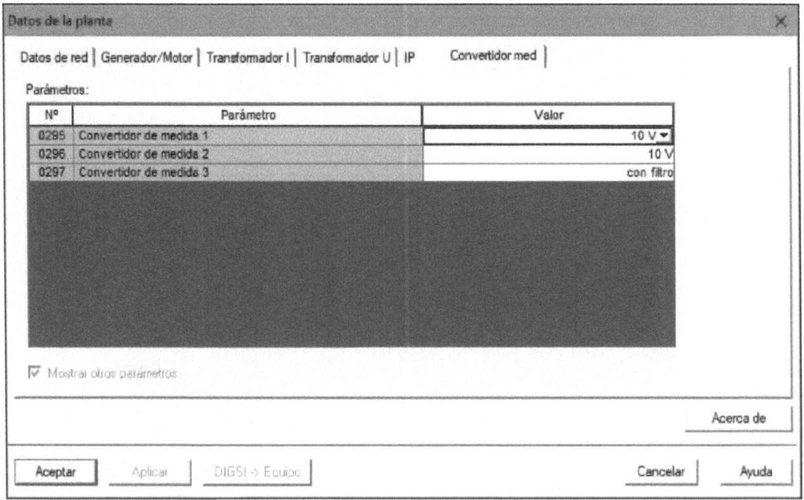

Measuring transducer 1 (K13 - K13) and measuring transducer 2 (K15 - K15) are used as the measuring transducer.

K16) will be used in Control Unit 7XT71(Sensitive Rotor Fault Protection with 1 to 3 Hz):

0295 Measurement converter 1: 10 V

0296 Measurement converter 2: 10 V

Measuring transducer 3 (K17-K18) will be used for the resistance unit.

3PP1326, for excitation voltage monitoring:

0297 Measurement converter 3: with filter

2. - Parameter Group A

2.1- S/I t. inv.(control/voltage depend.)

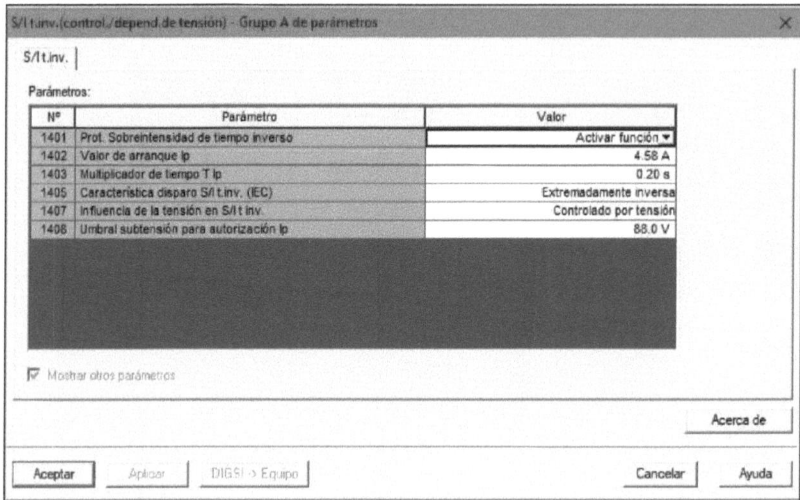

From generator electrical data:

Pn: 30 MW, Vn: 13.8 Kv, Fp: 0.8, In = 1590 Amp.

Ip Pickup = 1.15 x In = 1828 Amp Prim. = 4.57 Amp Secondary

Then:

1402 Ip start-up value: 4.57 A (in Secondary values)

1403 Time multiplier T Ip: 0.2 sec (Time Dial)

1405 Triggering characteristic S/I t inv.(IEC): Extremely Inverse

1407 Influence of voltage on S/I t inv: Voltage-controlled

1408 Undervoltage threshold for Ip. authorization: = 88 Volts (11040 Volts Prim.)

2.2.- Overload protection:

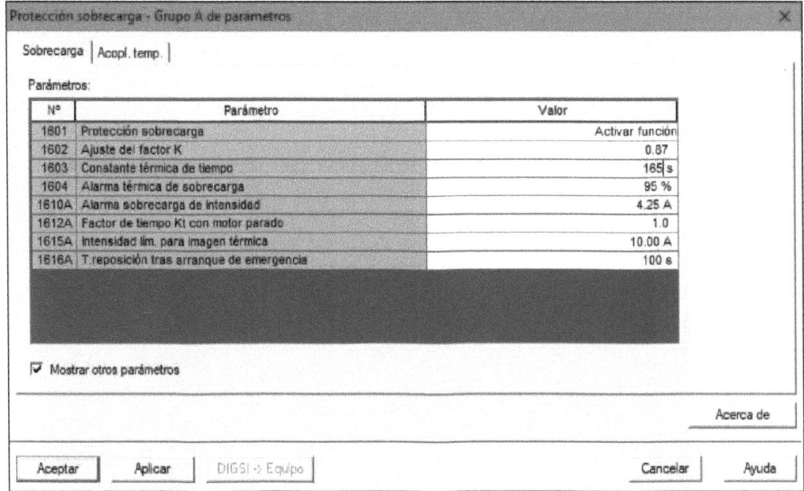

The K Factor = Imax prim / In Transf. prim = 1.1 In Gen / In Transf. prim = (1.1 x 1590) / 2000 = 0.8745.

then:

1602 K-factor adjustment: 0.87

Note: manufacturer's thermal characteristics are not available, therefore assumed:

Iload: 1.3xIn, Ttrip: 60 sec and Ipreload: 1xIn

$$\tau = \frac{t}{\ln \dfrac{\left(\dfrac{I_{Load}}{k \cdot I_n}\right)^2 - \left(\dfrac{I_{Preload}}{k \cdot I_n}\right)^2}{\left(\dfrac{I_{Load}}{k \cdot I_n}\right)^2 - 1}}$$

Then,

T = 60 sec / ln (1.4374) = 165 sec

Then the thermal time constant **1603:** 165 sec.

(To be verified with the characteristic tripping time of the overload protection).

1610A Current overload alarm: = 107 % = 1.07 x In = 1700 Amp Prim. = 4.25 Amp Sec.

1615A Current limit for thermal imaging: 252% = 2.52 x In = 10 Amp

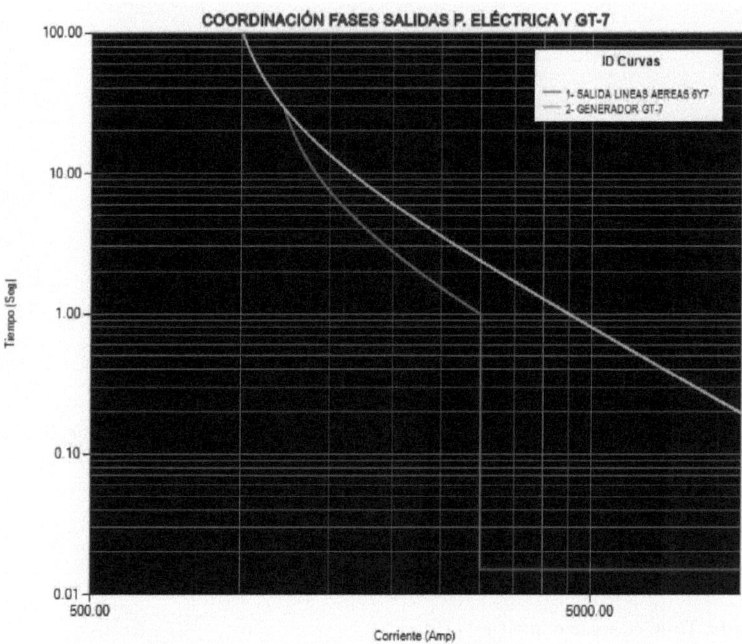

2.3 Load unbalance:

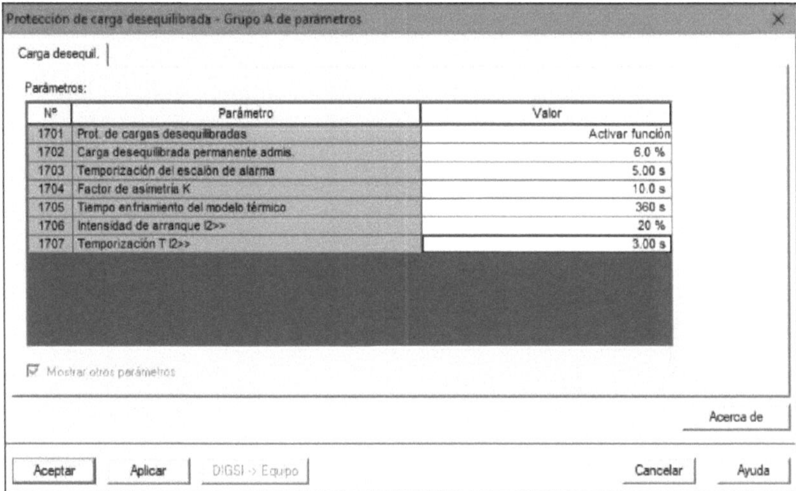

From generator electrical data:

Permissible continuous reverse current: 8 % In, then:

I2 = 8% (In Gen / In Transf. Prim) = 8 % (1590 / 2000) = 6.36 %.

1702 Permissible permanent unbalanced current I2: 6 %.

1703 Alarm step timing: 5 sec.

From the generator permissible reverse current diagram: $(I2^2\ t)$ = 10 sec, then:

1704 Asymmetry factor K: 10 sec.

The cooling time = $K / (I2)^2 = 10 / (0.08)^2 = 1562$ sec, then:

1705 Cooling time: 1562 sec.

1706 Start-up current I2 >>: 20%.

1707 Timing I2 >>: 3 sec.

Reverse current permissible diagram of the Generator GT-7

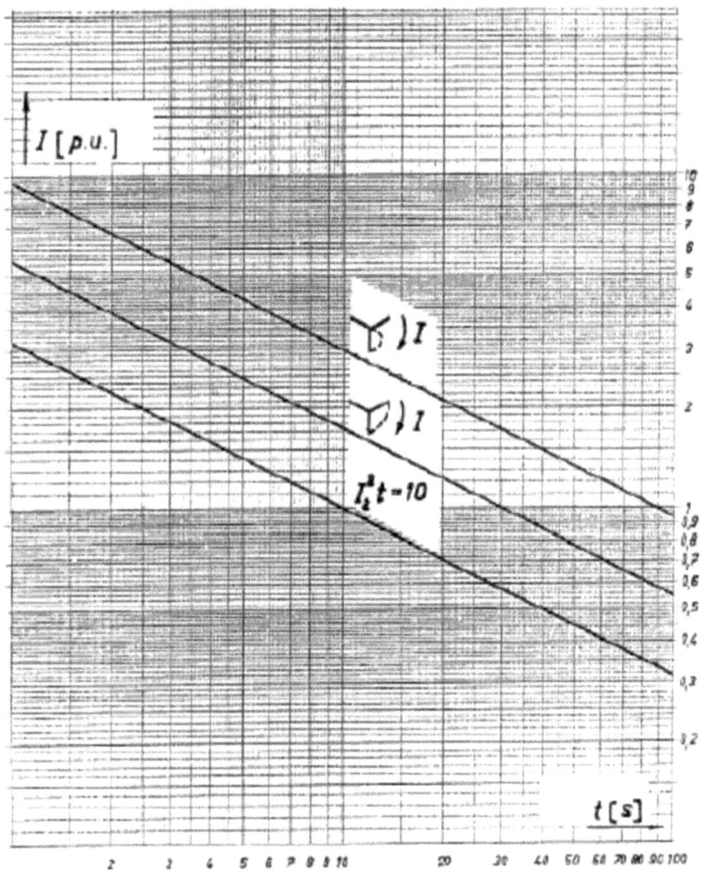

2.4.- Differential Protection

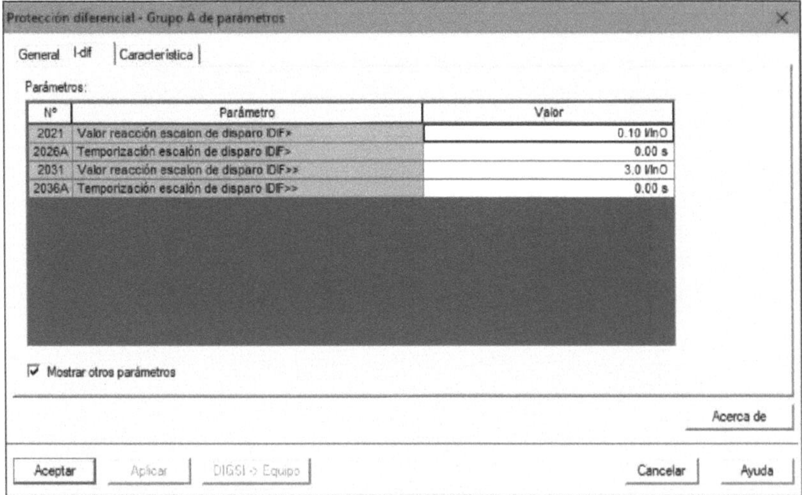

IDIFF>: 0.1I/In0 = 0.1 x In Gen = 0.1 x 1590 = 159 Amp

2021 Trigger step reaction value IDIF>: 0.1I/In0 **2026A** Trigger step timing IDIFF>: 0.00 sec.

IDIFF>>: 0.3I/In0 = 3 x In Gen = 3 x 1590 = 4770 Amp

2031 Trigger step reaction value IDIFF>>:: 3 I/In0 **2036A** Trigger step timing IDIFF>>: 0.00 Sec.

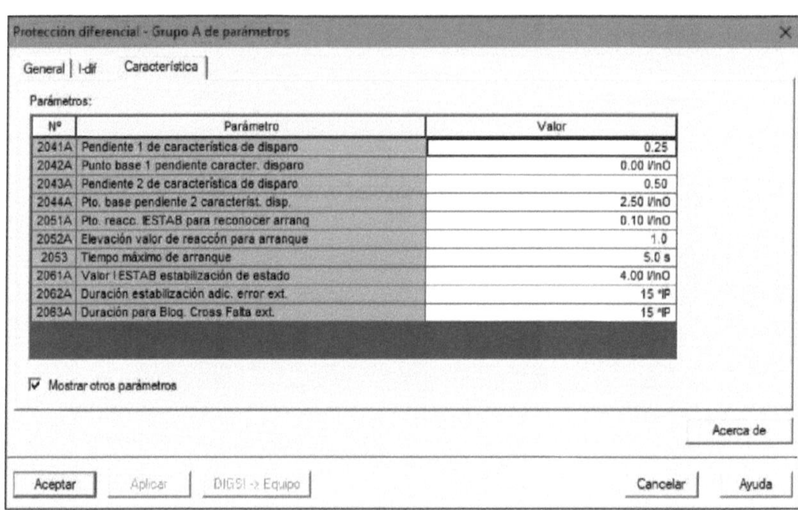

Operating characteristic of the generator differential protection:

Slope m1: 0.25, Slope m2: 0.50

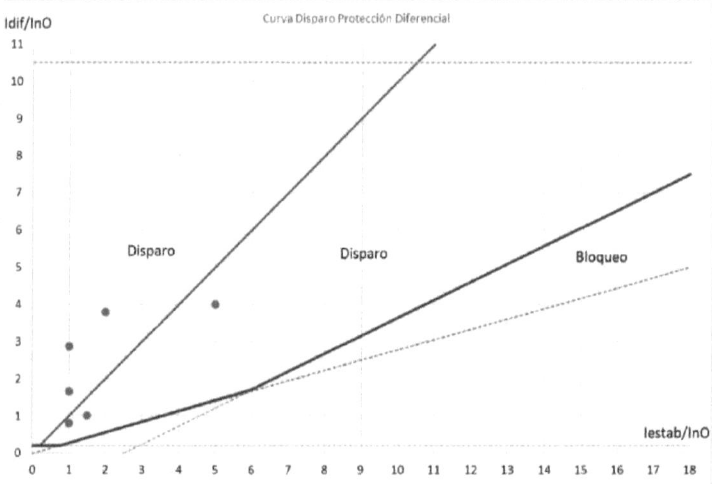

2.5.- Subexcitation

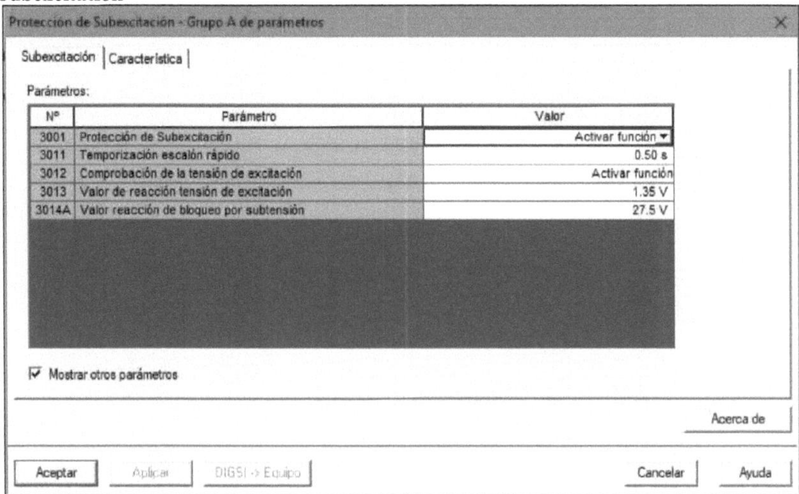

From the generator excitation data:

No Load voltage: ... UFo: 54 Volts, No Load current: ... IFo: 281 Amp

Rated voltage: UFn:195 Volts, Rated current:................IFn: 830 Amp

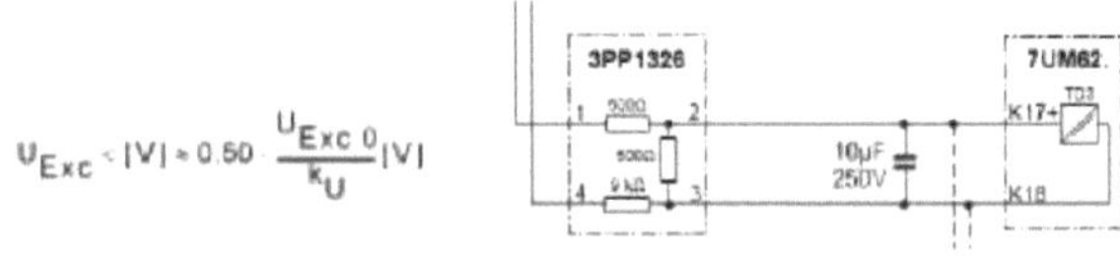

The ratio of the voltage divider 3PP1326: 20:1

Then:

Ku: (0.5+0.5+9)/ 0.5 = 20

UExc < = 0.5 x (54 / 20) Volts = 1.35 Volts

3013 Reaction value excitation voltage: 1.35 Volts

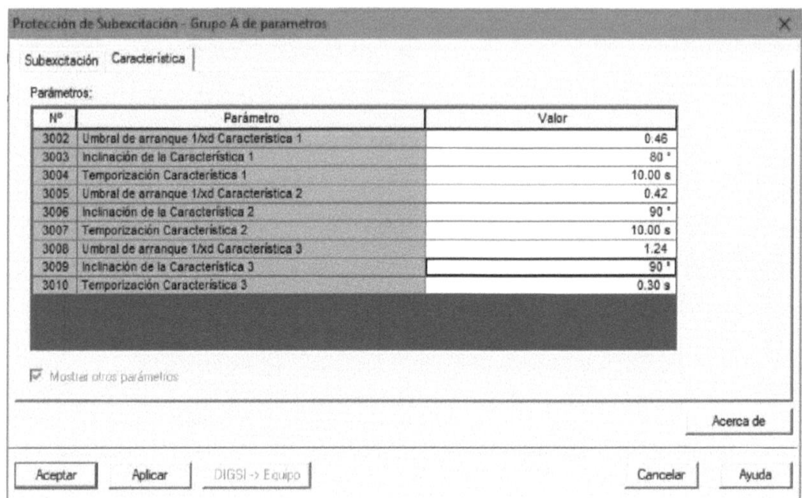

N°	Parámetro	Valor
3002	Umbral de arranque 1/xd Característica 1	0.46
3003	Inclinación de la Característica 1	80 °
3004	Temporización Característica 1	10.00 s
3005	Umbral de arranque 1/xd Característica 2	0.42
3006	Inclinación de la Característica 2	90 °
3007	Temporización Característica 2	10.00 s
3008	Umbral de arranque 1/xd Característica 3	1.24
3009	Inclinación de la Característica 3	90 °
3010	Temporización Característica 3	0.30 s

From generator electrical data:

Longitudinal synchronous reactance Xd = 228%.

Transient reactance X'd longitudinal = 19.5% Transient reactance X'd longitudinal = 19.5% Transient reactance X'd longitudinal = 19.5%

De: $(1 / xd) \times fs = (1 / 2.28) \times 1.05 = 0.46$

3002 Start-up threshold 1/xd Characteristic 1: 0.46

3003 Angle of inclination of feature 1: 80 ° (this must be between (60 and 80) °)

3004 Timing Characteristic 1: 10 sec.

De: $(1 / xd \text{ characteristic } 1) \times 0.9 = 0.46 \times 0.9 = 0.414$

3005 Start-up threshold 1/xd Characteristic 2: 0.42

3006 Angle of inclination of feature 2: 90°.

3007 Timing Characteristic 2: 10 sec.

From: $(xd + x'd) / 2 = (2.28 + 0.195) / 2 = 1.24$

3008 Start-up threshold 1/xd Characteristic 3: 1.24

3009 Angle of inclination of characteristic 3: 90 ° This should be between (80 and 110) °.

3010 Timing Characteristic 2: 0.3 sec.

Generator GT-7 load diagram

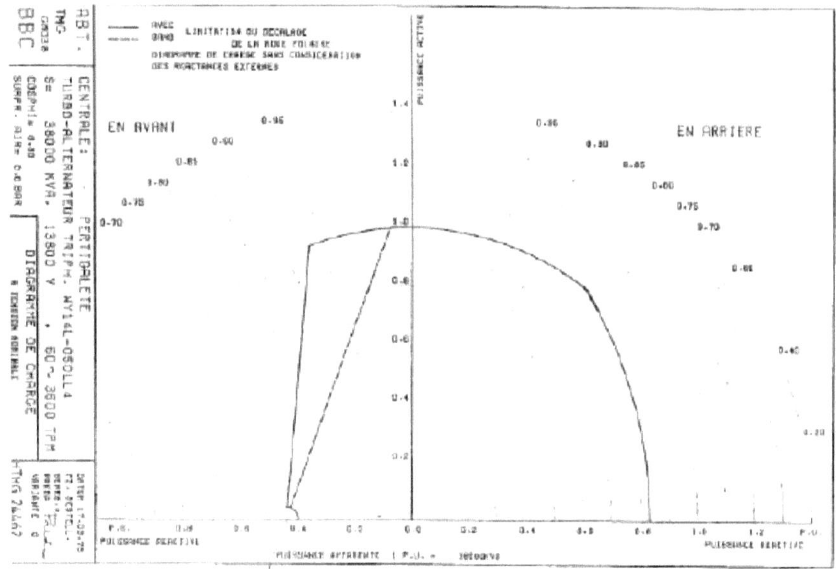

2.6. - Inverse Power

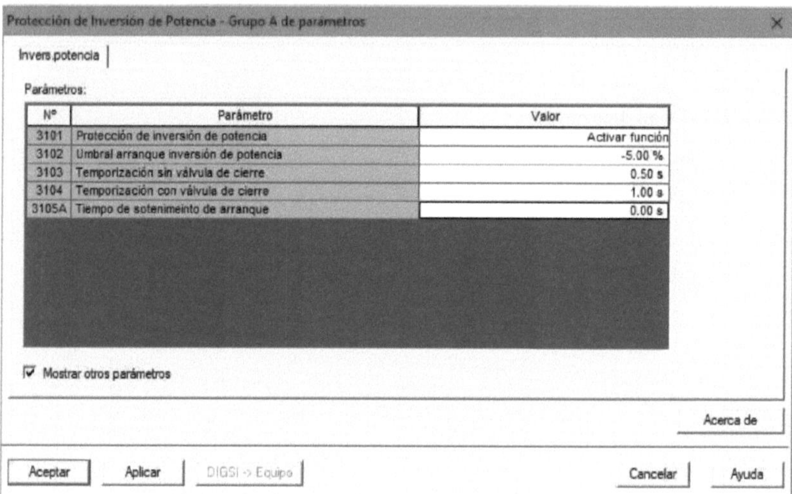

Recommended for gas turbines (3 to 5)% of rated power

3102 Power reversal start-up threshold: 5%.

3103 Timing (without shut-off valve): 0.5 Sec.

2.7.- Overvoltage:

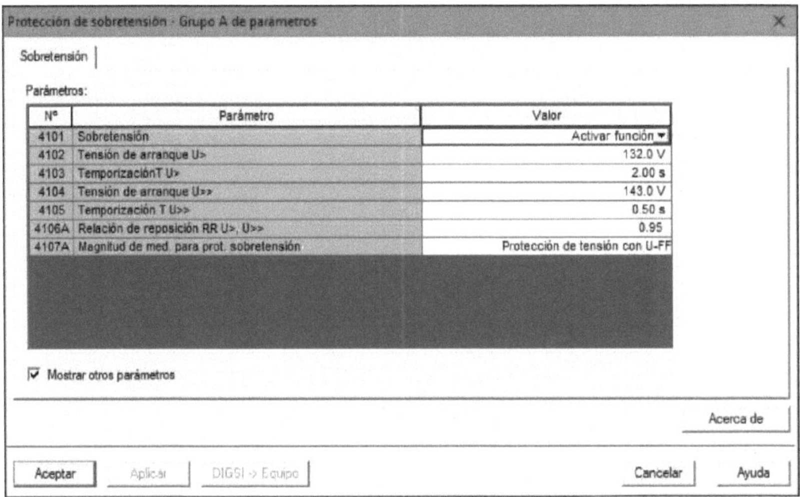

For U> = 1.2 x Un = 1.2 x 110 volts = 132 volts = 16.6 Kv

Then:

4102 U> Pickup: 132 V

4103 TU> Time Delay: 2 Sec.

For U>> = 1.3 x Un = 1.3 x 110 volts = 143 volts = 18 Kv

4104 U>> Pickup: 143 V

4105 TU>>Time Delay: 0.5 Sec.

2.8.- Frequency

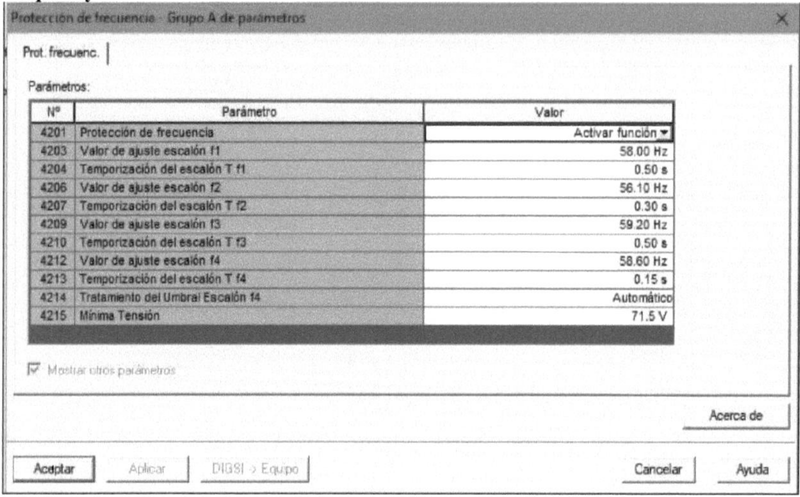

Based on the load and transient stability study of the cement plant:

4203   f1 Pickup: 58 Hz (generator stop)

4204   T f1 Time delay: 0.5 sec.

4206   f2 Pickup: 56.1 Hz (generator shutdown + gas turbine shutdown)

4207   T f2 Delay time: 0.3 sec.

4209   f3 Pickup: 59.2 Hz (Alarm #1)

4210   T f3 Delay time: 0.5 sec.

4212   f4 Pickup: 58.6 Hz (Alarm # 2)

4213   T f4 Delay time: 0.15 sec.

2.9.- Stator Ground Fault

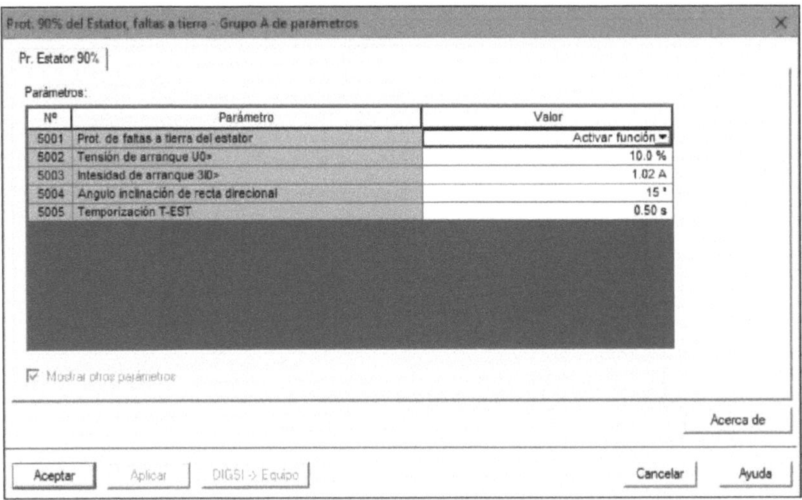

The plant's electrical system, being isolated to ground, has a directional ground fault relay operating in parallel to the grid. The selective stator ground leakage protection of the generators comprises a common ground fault relay type CUD (3xCUH90) per set of busbars. These detect the smallest neutral displacement as well as the phase affected by the fault. By means of a ground transformer (Open Delta) and a resistor, this relay causes the phase following the one affected by the leakage to be grounded. In addition, the CUD relay applies an appropriate phase-to-phase voltage to the ground fault relay type CG26i. The active current to ground produced by this change in the resistance of a phase, normally of the order of 10 A, flows in the branch belonging to the affected alternator and is therefore measured by its ground fault relay type CG26i. As soon as this relay is activated, it shuts down the alternator. This protection system can detect earth leakage over almost the entire length of the winding.

Ground fault on phase T, inside the protected zone:

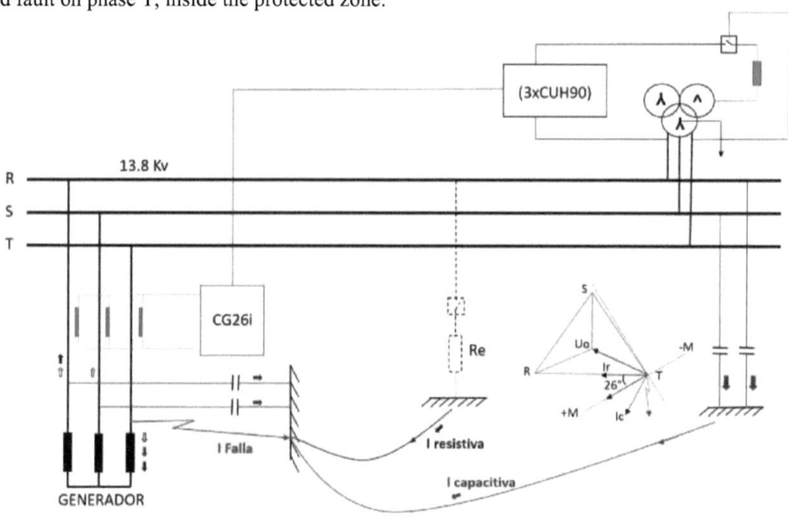

Ground fault on phase T, outside the protected zone:

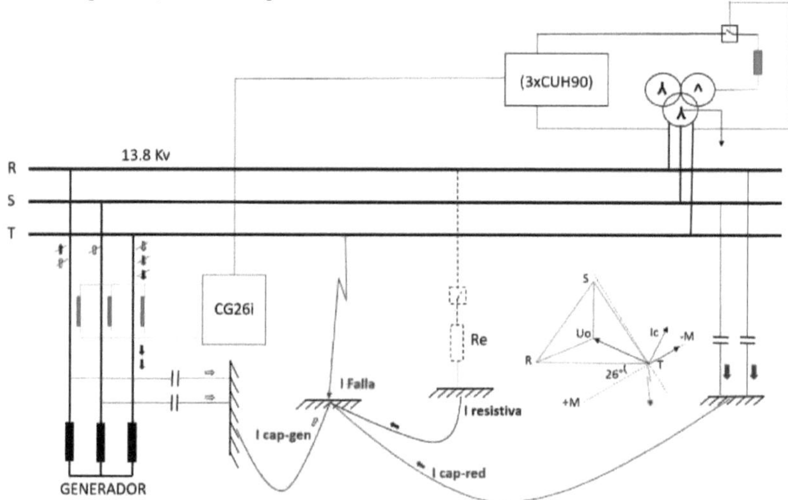

From the generator:

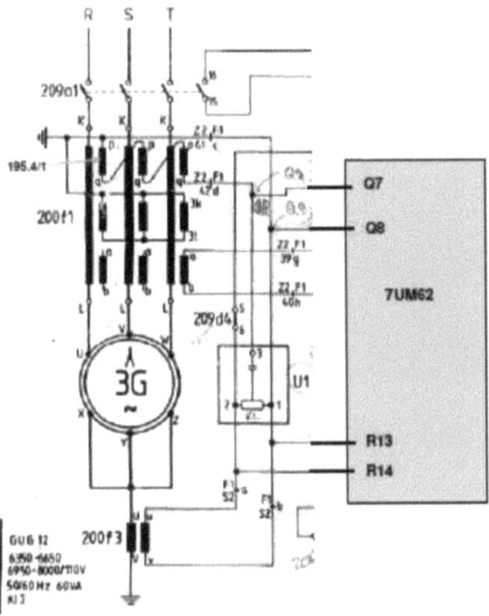

From transformer 200f3: Uo> = 10%, Ue = 11 Volts sec = 797 Volts prim.

Where:

$$U_{E0} = fe \times \left(\frac{UN_{Tprim}}{\sqrt{3}}\right) = \frac{13800}{\sqrt{3}} = 7967\ Volts = 7.97\ Kv$$

(fe = 1 in isolated network), then:

5002 Pickup voltage Uo> Pickup: 11 V

5003 3Io Pickup>: 1.02 Amp primary (to ensure higher sensitivity)

Angle **5004** for direction determination: 15 °.

5005 T-EST timing: 0.5 sec.

2.10.- Sensitive ground fault

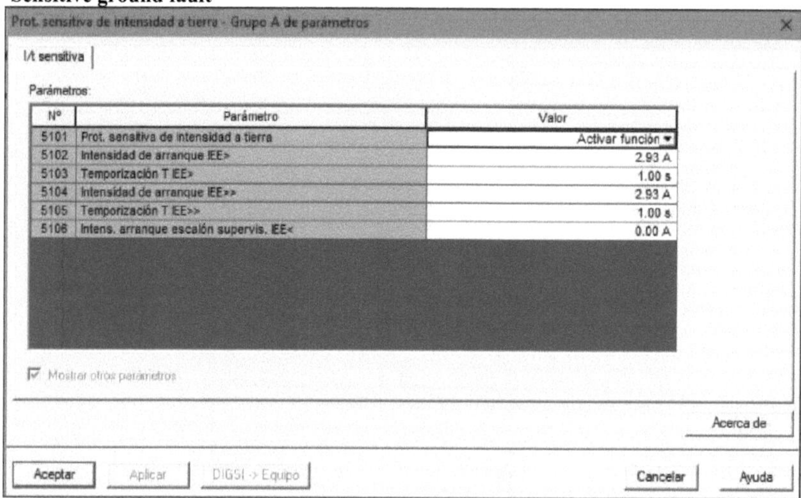

Prot. sensitiva de intensidad a tierra - Grupo A de parámetros

I/t sensitiva

Parámetros:

Nº	Parámetro	Valor
5101	Prot. sensitiva de intensidad a tierra	Activar función
5102	Intensidad de arranque IEE>	2.93 A
5103	Temporización T IEE>	1.00 s
5104	Intensidad de arranque IEE>>	2.93 A
5105	Temporización T IEE>>	1.00 s
5106	Intens. arranque escalón supervis. IEE<	0.00 A

Mostrar otros parámetros

Acerca de

Aceptar   Aplicar   DIGSI -> Equipo   Cancelar   Ayuda

This will be used as a backup for the "Stator Ground Fault Protection".

5102 IEE> Pickup: 3 amps, **5103** T IEE> Delay time: 1 sec.

2.11 Rotor ground fault

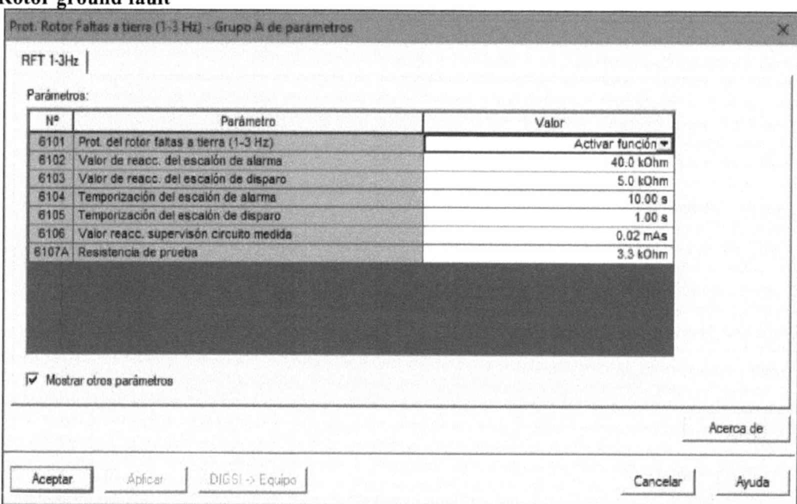

Prot. Rotor Faltas a tierra (1-3 Hz) - Grupo A de parámetros

RFT 1-3Hz

Parámetros:

Nº	Parámetro	Valor
6101	Prot. del rotor faltas a tierra (1-3 Hz)	Activar función
6102	Valor de reacc. del escalón de alarma	40.0 kOhm
6103	Valor de reacc. del escalón de disparo	5.0 kOhm
6104	Temporización del escalón de alarma	10.00 s
6105	Temporización del escalón de disparo	1.00 s
6106	Valor reacc. supervisón circuito medida	0.02 mAs
6107A	Resistencia de prueba	3.3 kOhm

Mostrar otros parámetros

Acerca de

Aceptar   Aplicar   DIGSI -> Equipo   Cancelar   Ayuda

This protection will only be used as an alarm, it will not stop the generator.

Configuration used:

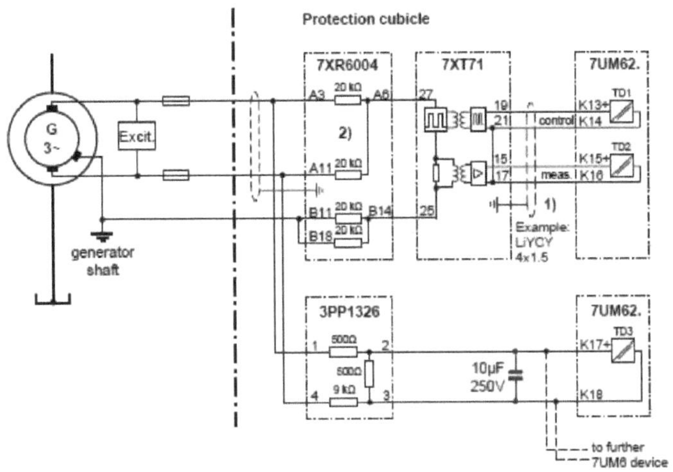

Binary Outputs (SB) of the protection relay

- Circuit Breaker Generator (212d): SB1, SB7

- Field Breaker (210a): SB2

- Disturbance relay (211f): SB5

- General alarm (211e): SB4

- 00511 Relay Trip (AS): SB10 (For current injection tests)

- arrq Uo> (AI): SB14

- arrq Io> (AI): SB15

- 05403 RFT 1-3 Hz Warm (AS): SB16

- 05403 RFT 1-3 Hz Trip (AS): SB16

- Alarm blink f/t rotor open (AI): SB17

3.   - Alarm signals and triggers

LED		PROTECCIÓN	description
1		Disp. ΔI >	differential protection
2		Disp. I>/U<	voltage-controlled overcurrent trip
3		Alarm I²t	overload alarm
4		Disp. P<	reverse power trip
5		Disp. U>	overvoltage trip
6		Disp. 1/Xd	failure excitation
7		Alarm I2>	current unbalance alarm
8		Disp I2>	tripping due to current unbalance
9		f< 59.2 Hz	low frequency alarm at 59.2 hz
10		f< 58.6 Hz	low frequency alarm at 58.6 hz
11		f< 56.1 Hz	gen shot + your queen by low frequency at 56.1 hz
12		f< 58.0 Hz	low frequency generator trip at5b.0hz
13		Disp Iee> DIR	directional ground fault uo>+3lo>+ang >15*
14		Disp IEE>	ground fault not to be used as backup
15		Alarm Uo>	ground voltage performance
16		Alarm Io>	current to ground actuation
17		Disp Re<	rotor ground fault
18		Re< OPEN	open ground fault circuit rotor

Continuos Function Chart (CFC)" tool

All alarm signals will be programmed to flash and four outputs will be added to supply signals to additional lamps.

For alarm signals, we built a logic block using: RS Flipflop block + a Blink module.

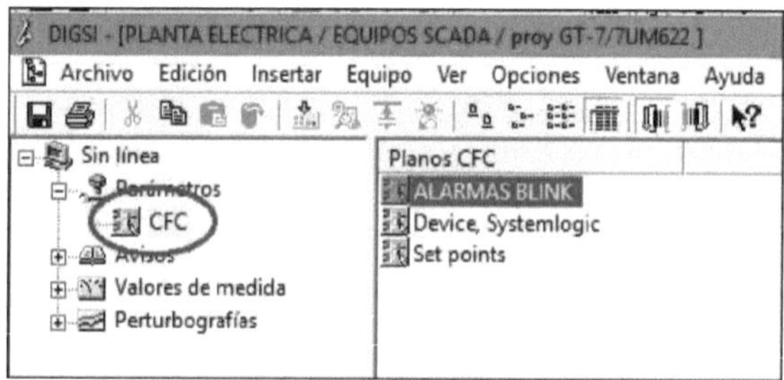

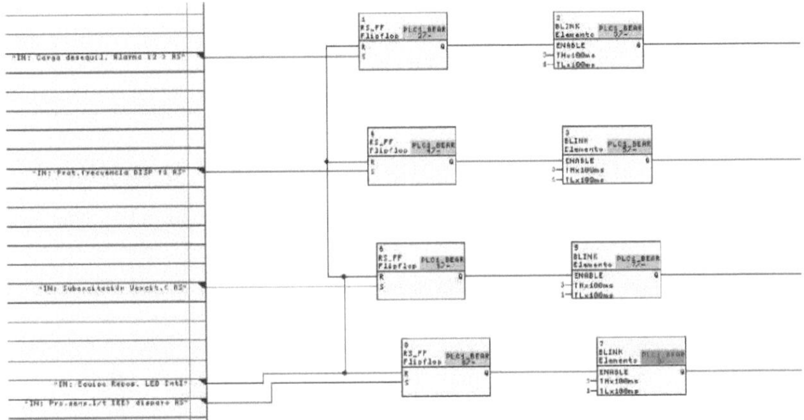

Photos of assembly and installation

Relay in Panel Surface Mounting

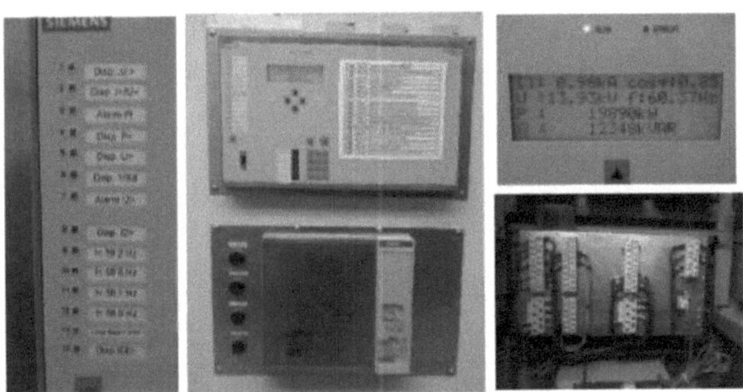

Relay in Panel Surface Mounting

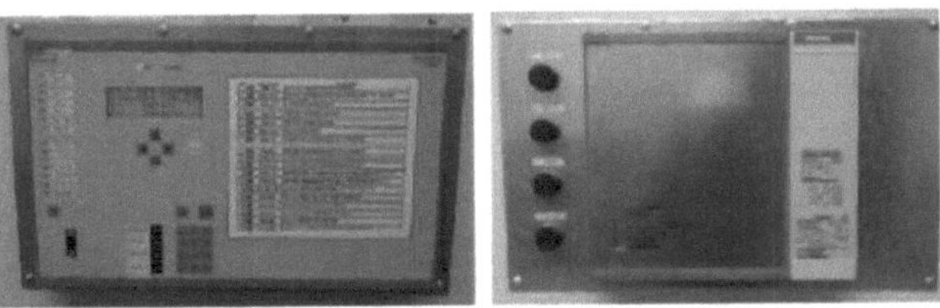

4.- Primary tests

For the primary test a PROGRAMMA - FREJA 300 relay test system and FLUKE 289 RMS Multimeter are used.

4.1.- Rotor ground fault:

1 - Machine stop with coals inserted into rotor

Number	Description	Value
00755	f gen	1.4 Hz
00757	U gen	50.3 V
00758	I gen	0.00 mA
00759	Qc	0.042 mAs
00761	Rerde	999.9 kOhms

2. - Machine stop with coals outside the rotor

Number	Description	Value
00755	f gen	1.4 Hz
00757	U gen	50.3 V
00758	I gen	0.00 mA
00759	Qc	**0.010 mAs**
00761	Rerde	999.9 kOhms

3,- Machine stop with grounded rotor

Number	Description	Value
00755	f gen	1.4 Hz
00757	U gen	50.3 V
00758	I gen	**2.43 mA**
00759	Qc	**0.835 ~ 0.777 mAs**
00761	Rerde	**0.00 kOhms**

4 - Machine stop with circuit interruption (input 27 from 7XT71)

Number	Description	Value
00755	f gen	1.4 Hz
00757	U gen	50.3 V
00758	I gen	0.00 mA
00759	Qc	**0.003 mAs**
00761	Rerde	kOhms

Pertubography:

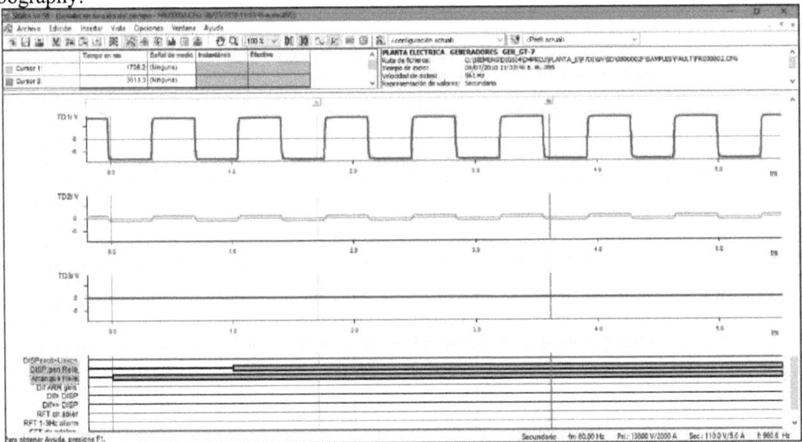

Events:

Número	Aviso	Valor	Fecha y hora
00301	Falta en Red, numerado	1 - Entra	08.07.2010 11:33:46.296
00302	Perturbación,evento de faltas	1 - Entra	08.07.2010 11:33:46.296
00501	Arranque general del relé de protección	Entra	0 ms
05406	Prot. rotor f/t (1-3 Hz) Arranque Re<<	Entra	0 ms
00511	Disparo del relé (general)	Entra	999 ms
05407	Prot. rotor f/t (1-3 Hz) Disparo Re<<	Entra	999 ms
00576	Intens. de desconexión (prim.) L1 Lado 1	0.00 kA	1016 ms
00577	Intens. de desconexión (prim.) L2 Lado 1	0.00 kA	1016 ms
00578	Intens. de desconexión (prim.) L3 Lado 1	0.00 kA	1016 ms
00579	Intens. de desconexión (prim.) L1 Lado 2	0.00 kA	1016 ms
00580	Intens. de desconexión (prim.) L2 Lado 2	0.00 kA	1016 ms
00581	Intens. de desconexión (prim.) L3 Lado 2	0.00 kA	1016 ms
05012	Tensión UL1E en el disparo	0.00 kV	1016 ms
05013	Tensión UL2E en el disparo	0.00 kV	1016 ms
05014	Tensión UL3E en el disparo	0.00 kV	1016 ms
05015	Potencia activa P en el disparo	0.00 MW	1016 ms
05016	Potencia reactiva Q en el disparo	0.00 MVAR	1016 ms
05017	Frecuencia en el disparo	0.00 Hz	1016 ms
05701	Int.dif.L1 con disp.s. ret (onda fundam)	0.00 I/In0	1016 ms
05702	Int.dif.L2 con disp.s. ret (onda fundam)	0.00 I/In0	1016 ms
05703	Int.dif.L3 con disp.s. ret (onda fundam)	0.00 I/In0	1016 ms
05704	Int. est.L1 con disp.s.ret (val.rectif.)	0.00 I/In0	1016 ms
05705	Int. est.L2 con disp.s.ret (val.rectif.)	0.00 I/In0	1016 ms
05706	Int. est.L3 con disp.s.ret (val.rectif.)	0.00 I/In0	1016 ms
05406	Prot. rotor f/t (1-3 Hz) Arranque Re<<	Sale	38017 ms
00301	Falta en Red, numerado	1 - Sale	08.07.2010 11:34:24.314

4.2.- Stator Ground Fault (Test # 1)

Vectorial Chart:

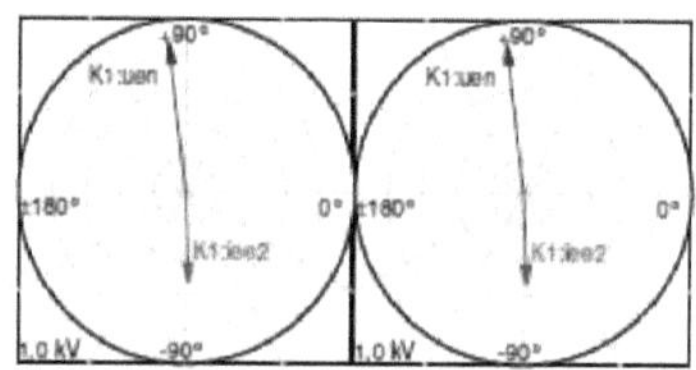

	t en ms	Señal de medida	Dimensión	Fase	Real	Imag
Cursor 1:	645,2	K1:uan	0,88 kV	97,2°	-0,11 kV	0,87 kV
Cursor 2:	646,3	K1:iee2	5,53 A	-87,3°	0,26 A	-5,52 A
C2 - C1	1,1	K1:iee2 - K1:uan		-184,6°		

Pertubography:

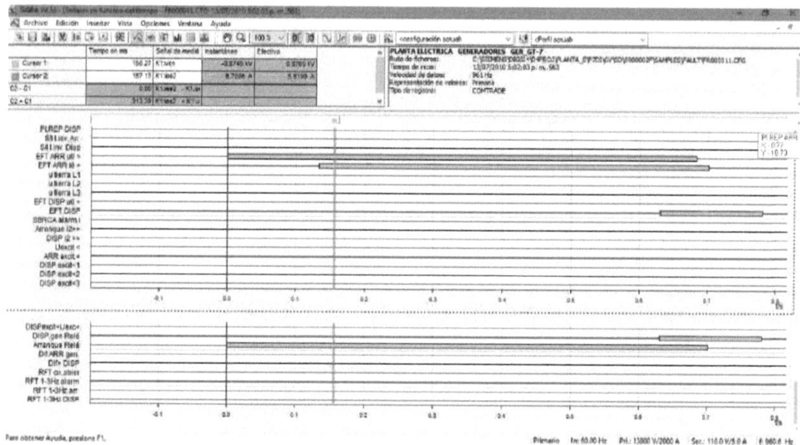

4.3.- Stator Ground Fault (Fault # 2)

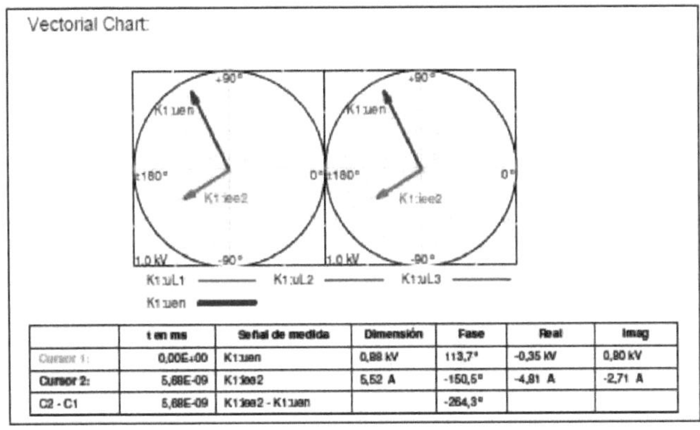

	t en ms	Señal de medida	Dimensión	Fase	Real	Imag
Cursor 1:	0,00E+00	K1:uen	0,88 kV	113,7°	-0,35 kV	0,80 kV
Cursor 2:	5,68E-09	K1:iee2	5,52 A	-150,5°	-4,81 A	-2,71 A
C2 - C1	5,68E-09	K1:iee2 - K1:uen		-264,3°		

Pertubografía:

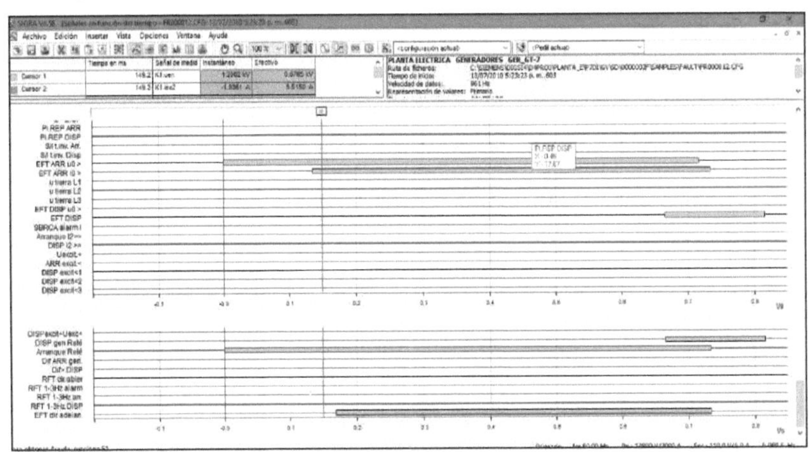

4.4.- Stator Ground Fault (Fault # 3)

Vectorial Chart:

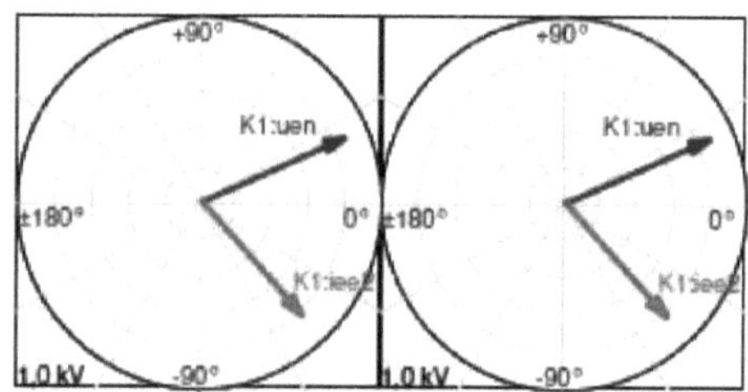

	t en ms	Señal de medida	Dimensión	Fase	Real	Imag
Cursor 1:	-1,73E-09	K1:uen	0,68 kV	23,8°	0,80 kV	0,35 kV
Cursor 2:	0,00E+00	K1:iee2	6,47 A	-47,6°	5,71 A	-6,26 A
C2 - C1	1,73E-09	K1:iee2 - K1:uen		-71,4°		

Oscillographic Record:

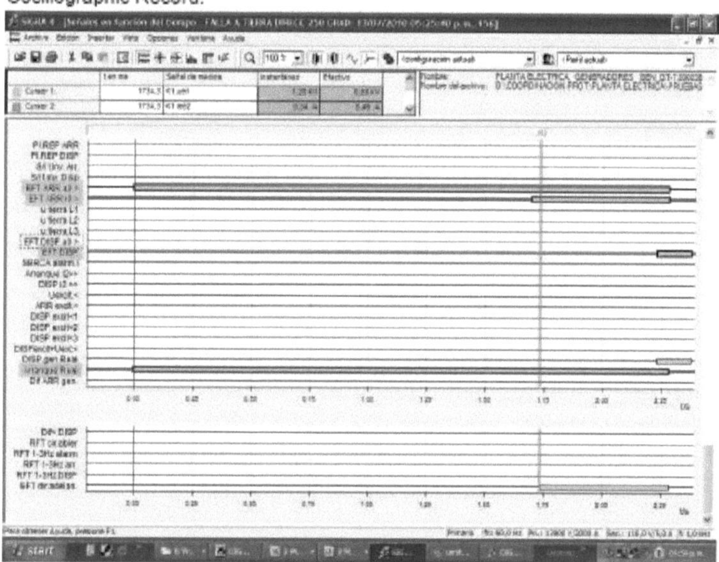

38

4.5.- Triggering Prot. Differential

Oscillographic Record:

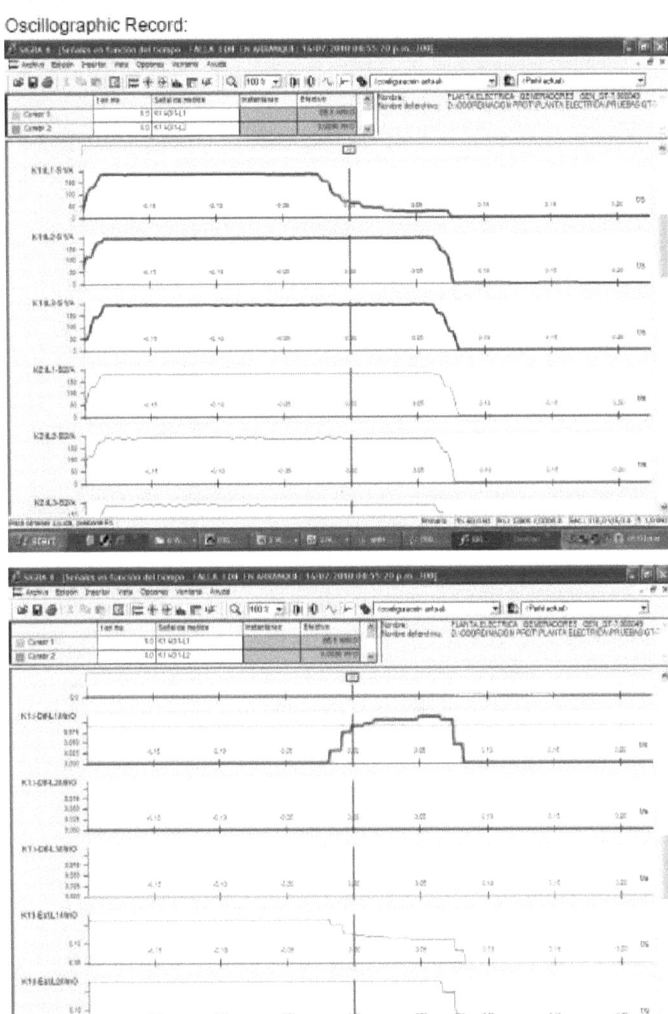

Diff> Trip: Oscillographic Record:

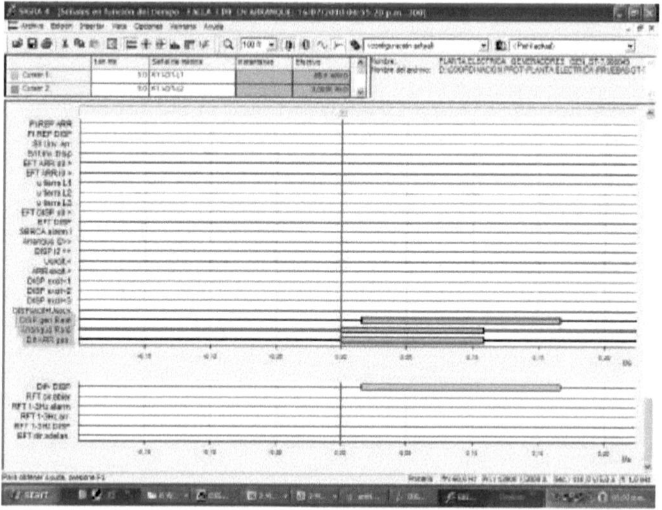

5.- Shots and Events

The generator started on 07/16/2010, these have been the events and faults recorded by the new relay 7UM62.of the GT-7 generator:

Failure Event #1

Date: 08/09/2010, Time: 01:16 pm

Fault description: Overcurrent on transmission line # 8 Protection trip (I>) of relay 7SD610 on the generation side Fault currents: IL1: 5.21 KA, IL2: 5.56 KA, IL3: 5.36 KA, Single line diagram of the system at the time of the fault:

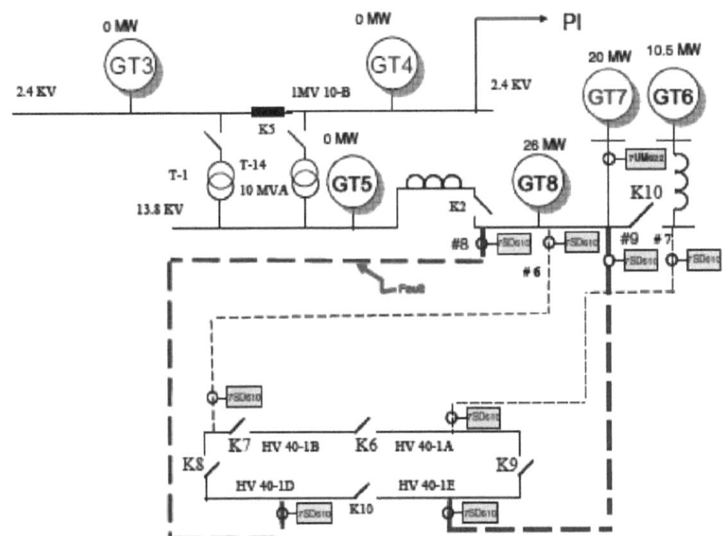

Oscillography and events displayed on relay 7UM622 of the GT-7 generator:

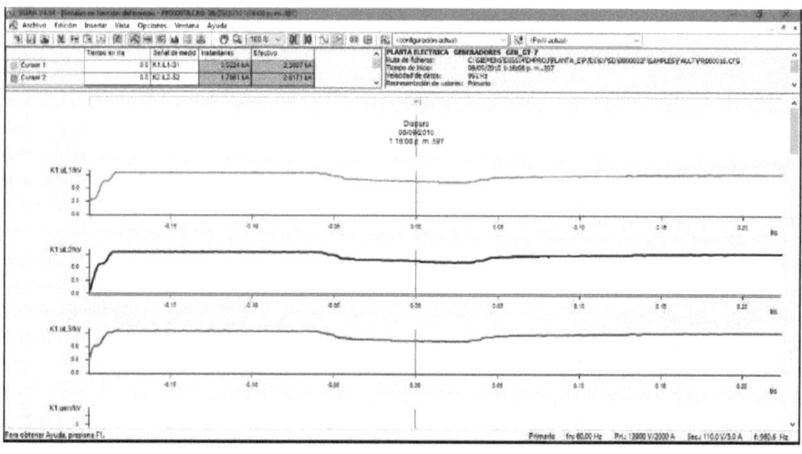

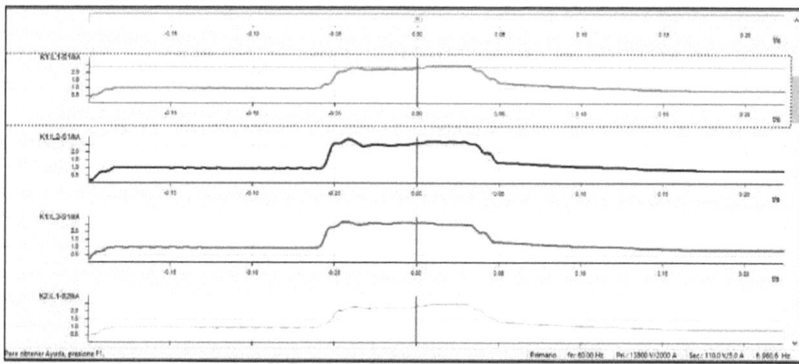

DIGSI - [Avisos de perturbación - 000062 / 08/09/2010 1:16:08.597 p. m. - PLANTA ELECTRICA / GENERADORES / GEN_GT-7/7UM622]

Archivo   Edición   Insertar   Equipo   Ver   Opciones   Ventana   Ayuda

Número	Aviso	Valor	Fecha y hora	Causante
00301	Falta en Red, numerado	62 - Entra	08.09.2010 13:16:08.597	
00302	Perturbación.evento de faltas	62 - Entra	08.09.2010 13:16:08.597	
00501	Arranque general del relé de protección	Entra	0 ms	
01897	Arranque S/I t.inv. Fase L2	Entra	0 ms	
01898	Arranque S/I t.inv. Fase L3	Entra	0 ms	
01899	Arranque S/I t.inv.	Entra	0 ms	
05165	Prot. carga deseq. arranque I2>	Entra	0 ms	
01896	Arranque S/I t.inv. Fase L1	Entra	16 ms	
05631	Protección diferencial arranque general	Entra	16 ms	
01896	Arranque S/I t.inv. Fase L1	Sale	50 ms	
01897	Arranque S/I t.inv. Fase L2	Sale	50 ms	
01898	Arranque S/I t.inv. Fase L3	Sale	50 ms	
01899	Arranque S/I t.inv.	Sale	50 ms	
05165	Prot. carga deseq. arranque I2>	Sale	66 ms	
05631	Protección diferencial arranque general	Sale	108 ms	
00301	Falta en Red, numerado	62 - Sale	08.09.2010 13:16:08.706	

Failure event #2

Date: 02/27/2012, Time: 05:07 pm

Background:

The GT-7 generator, had a shutdown for repair by technical personnel from ALSTON - Switzerland, the repair work started on 01/16/2012 and consisted of:

Replacement of the damaged winding bars at the star point.

Fixing of loose wedges and injection of DVV resin.

Once the repairs are completed, we proceed with the no-load start-up, presenting a ground fault at the star point.

connection bars in the star point before earth fault   connection bars in the star point after earth fault

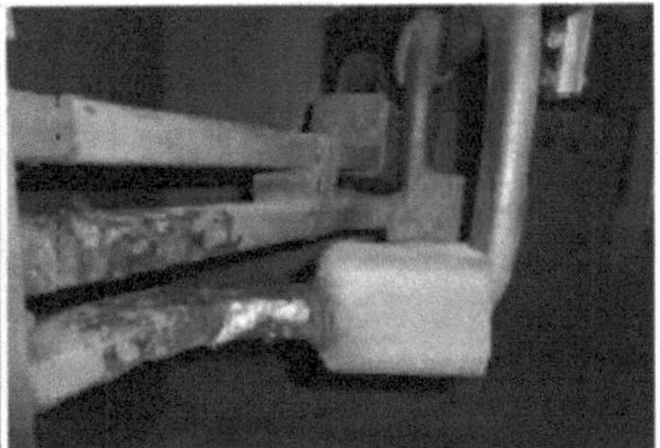

detail of the bar failed in the star point

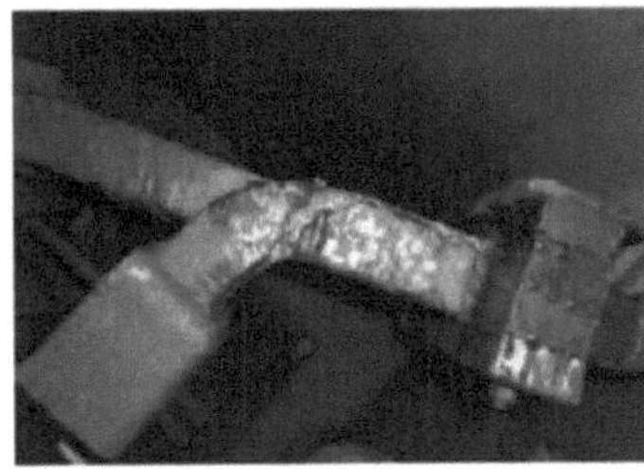
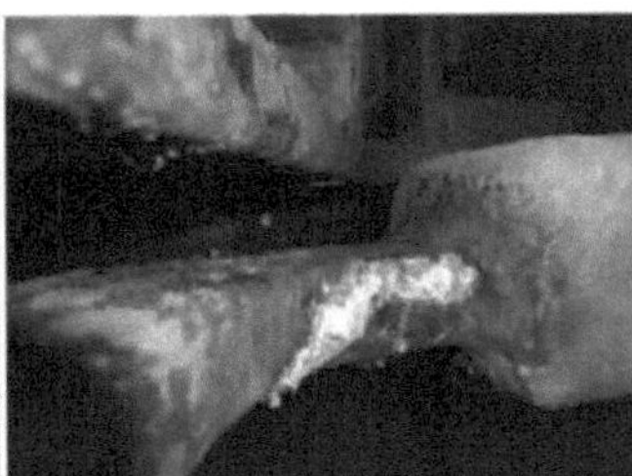

The protection relay 7UM622 operated immediately, actuating the stator ground fault and sensitive earth fault function IEE>.

Events:

Archivo   Edición   Insertar   Equipo   Ver   Opciones   Ventana   Ayuda

Número	Aviso	Valor	Fecha y hora
00301	Falta en Red, numerado	21 - Entra	27.02.2012 17:07:48.495
00302	Perturbación,evento de faltas	21 - Entra	27.02.2012 17:07:48.495
00501	Arranque general del relé de protección	Entra	0 ms
05186	Prot.estator f/t (U0>) arranque	Entra	0 ms
05334	Prot.subexcit. bloqueo por subtensión	Sale	299 ms
05188	Prot.estator f/t (I0>) arranque	Entra	850 ms
05214	Prot.frecuencia bloqueo por subtensión	Sale	916 ms
01224	Arranque escalón IEE> detección tierra	Entra	1350 ms
01221	Arranque escalón IEE>> detección tierra	Entra	1350 ms
06568	Arranque prot.de tensión, Esc U>	Entra	2017 ms
00511	Disparo del relé (general)	Entra	2348 ms
01226	Disparo protección-IEE IEE>	Entra	2348 ms
01223	Disparo protección-EEE IEE>>	Entra	2348 ms
00576	Intens. de desconexión (prim.) L1 Lado 1	0.00 kA	2367 ms
00577	Intens. de desconexión (prim.) L2 Lado 1	0.00 kA	2367 ms
00578	Intens. de desconexión (prim.) L3 Lado 1	0.00 kA	2367 ms
00579	Intens. de desconexión (prim.) L1 Lado 2	0.00 kA	2367 ms
00580	Intens. de desconexión (prim.) L2 Lado 2	0.00 kA	2367 ms
00581	Intens. de desconexión (prim.) L3 Lado 2	0.01 kA	2367 ms
05012	Tensión UL1E en el disparo	9.66 kV	2367 ms
05013	Tensión UL2E en el disparo	9.66 kV	2367 ms
05014	Tensión UL3E en el disparo	9.66 kV	2367 ms
05015	Potencia activa P en el disparo	0.00 MW	2367 ms
05016	Potencia reactiva Q en el disparo	-0.10 MVAR	2367 ms
05017	Frecuencia en el disparo	60.09 Hz	2367 ms
05701	Int.dif.L1 con disp.s. ret (onda fundam)	0.00 I/In0	2367 ms
05702	Int.dif.L2 con disp.s. ret (onda fundam)	0.00 I/In0	2367 ms
05703	Int.dif.L3 con disp.s. ret (onda fundam)	0.00 I/In0	2367 ms
05704	Int. est.L1 con disp.s.ret (val.rectif.)	0.00 I/In0	2367 ms
05705	Int. est.L2 con disp.s.ret (val.rectif.)	0.00 I/In0	2367 ms
05706	Int. est.L3 con disp.s.ret (val.rectif.)	0.00 I/In0	2367 ms
05336	Prot.subexcit.tensión de excit.muy baja	Entra	2433 ms
06568	Arranque prot.de tensión, Esc U>	Sale	2550 ms
01224	Arranque escalón IEE> detección tierra	Sale	3183 ms
01221	Arranque escalón IEE>> detección tierra	Sale	3183 ms
05214	Prot.frecuencia bloqueo por subtensión	Entra	4066 ms
05188	Prot.estator f/t (I0>) arranque	Sale	5367 ms
05334	Prot.subexcit. bloqueo por subtensión	Entra	7399 ms
05186	Prot.estator f/t (U0>) arranque	Sale	13017 ms
00301	Falta en Red, numerado	21 - Sale	27.02.2012 17:08:01.512

Vector diagram of the failure:

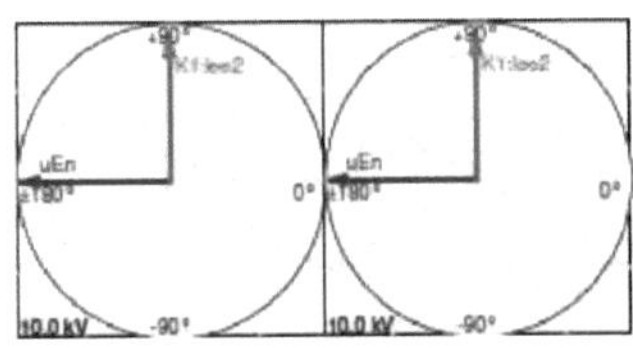

Pertubography:

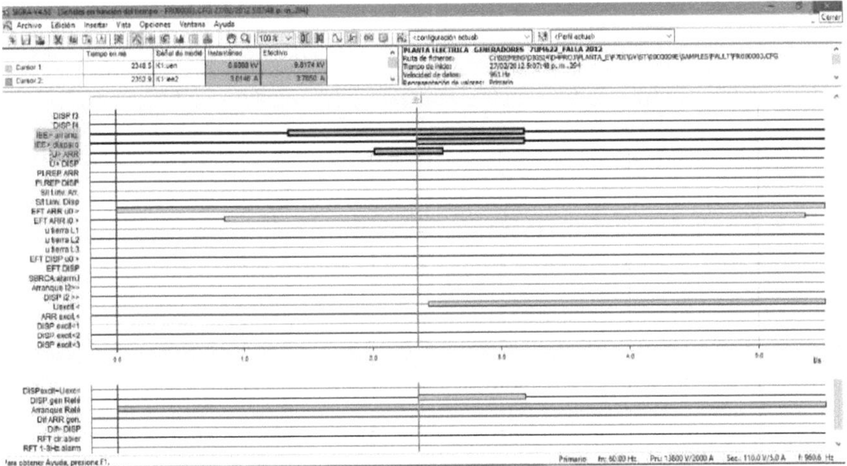

Bibliography:

[1] Multifunctional Machine Guard 7UM62 V4.6. Siemens AG 2010. Manual order No. C53000-G1178-C149-2.

[2] Optimum Motor Protection with SIPROTEC Protection Relays. Siemens AG 2006. Order No. E50001-K4454-A101-A1-7600.

[3] Machine Protection Setting Exercises. Siemens AG.

[4] Herrmann, Hans-Joachim: Generator Protection Rotor -Earth-Fault Protection, Siemens AG.

[5] Study and Calculation Data of Ground Fault Protection in 15 KV Power Supply, Mitsubishi Corporation 1989.

[6] Pertigalete Network Study Venezolana de Cementos, Brown Boveri - Werke. 1984

I want morebooks!

Buy your books fast and straightforward online - at one of world's fastest growing online book stores! Environmentally sound due to Print-on-Demand technologies.

Buy your books online at
www.morebooks.shop

Kaufen Sie Ihre Bücher schnell und unkompliziert online – auf einer der am schnellsten wachsenden Buchhandelsplattformen weltweit! Dank Print-On-Demand umwelt- und ressourcenschonend produziert.

Bücher schneller online kaufen
www.morebooks.shop